SHORE LINE
an imprint of WIDE OPEN SEA

First published in paperback and ebook 2020

Wide Open Sea · Art · Heritage · Press
hello@wideopensea.co.uk
wideopensea.co.uk

ISBN 978 1 9161148 52
A CIP catalogue record for this book is available from the
British Library.

Cover and endpage illustrations by Suzy Thomas.

KEEPING HENS
A Chatty Guide to Chickens

by
Jane Furnival

with Andy and Will Tribble

SHORE LINE
inspiring encounters

Contents

Foreword
by Will Tribble

Hello. I'm Will Tribble, the eldest son of Jane Furnival and Andrew Tribble, who wrote most of this book.

If you have purchased a book called *Keeping Hens* solely for information about keeping hens, all you need to know are two things. The first is that most of this book was written between 2010 and 2012, but for reasons explained below it was published about ten years later. The second is that it's based on small scale personal experiences, written by and for people keeping hens in their garden (or field) rather than industrial scale farming, although I hope it's useful for everyone.

As such, I'm going to occasionally butt into the main body of the text with updates, clarifications and disclaimers from the distant future of 2020, based on the advice of people who know a lot more about chickens than I do. I'll do my best not to upset my parents' writing.

Speaking of experts, I'd like to thank Gaynor Davies of the British Hen Welfare Trust, Francine Raymond, Terry Dunk and John Harrison, for replying to my emails with lots of helpful advice and suggestions. As Francine pointed out, 'there are as many ways of looking after chickens as there are childrearing philosophies,' but I hope it gives you a rounded idea of things that can (and can't) be done. In particular there's some big changes in hen feeding advice that I'll deal with later in the book.

Jane Furnival was a firework. An intensely driven and resourceful person you would always remember meeting. We didn't have much money when I was born, and so she started collecting money saving tips; she turned these tips into a newspaper column, then a book, then a TV series.

My parents eventually bought a beautiful house in Cheam Village called The Old Rectory, which she immediately rented out for weddings and photo shoots and ghost hunts, with the whole family thrown in as butlers, runners, or paranormal investigators. Now that we had the space, mum could live her suburban farming dream of keeping hens.

I think mum partly became interested in henkeeping out of a desire to be self-sufficient. We also had a vegetable patch, and if we could have logistically kept cows I think she might have tried. As she lays out (hahaaaa) later, a few well-looked-after chickens can easily lay enough eggs for you and still have plenty to spare.

But really I think mum wanted hens because she loved animals. She loved raising them, watching them pottering about in the garden or sneaking where they shouldn't, loved having dignified conversations with them.

Having a space filled with colourful, comical, clucking life is as good a reason as any for keeping hens. But you must be prepared to care for them, as she did. We have many memories of her sitting in the darkened living room, TV flickering near silently, a sick hen wrapped in towels in her arms, as she tried to nurse it through the night. If a hen was missing, we would all be out hunting for it, long after dark.

As a family we supported her in all of this. My father designed and constructed from scratch an elaborate shed-sized henhouse, custom to our specific needs, using nothing but some library research and a few trips to the DIY shop. He has fully hand-diagrammed this process later in the book. There are a few coffee stains but they don't get in the way. My brother Henry has digitally recreated our chickenshed to give you a full, clear view of the whole building, alongside photos of the real thing.

There are also photos of our chickens, a couple of us, and our dog Boo. Boo was a beautiful, loveable dog, the size of a man in a very furry suit, and in many way my mother's fourth son. Mum once won a prize for owners that looked most like their dogs and was delighted. After a few bad starts and, I need to emphasise for the sake of your hens, a LOT of home training, Boo became an excellent hen guardian. He would run round the house barking at the first hint of sundown, excitedly rounding us up so he in turn could round up the hens. He would delightedly bound around the garden, rooting through every bush, pulling us behind him down roads and sidestreets if he needed to, until suddenly a hen would emerge clucking from a shrubbery and we'd do our best to get it home.

We all tried our best to herd the flock, learnt to keep one eye on the window, adapted to a routine that frequently included a clucking playful Benny Hill chase routine each sundown. The hens, at least, understood that while they would like to pull up the carrots, it was probably for the best that I was freeing their trapped claws from the netting and gently carrying them out of the vegetable patch. The cockerels were a different story. I remember cockerels that at less than a year old were somehow filled with aged and righteous *how dare you* fury, staring me down with Peter Capaldi eyes as I awkwardly tried to herd them to their beds with a flapping umbrella. Velociraptors apparently were a lot like chickens; having turned away from a cockerel on the other side of the garden, only to turn back seconds later and see him, claws outstretched, skittering to a halt next to me, I can believe it. And while I am against the fetishized upper class notions of foxhunting, henkeeping has given me an abiding Boggis Bunce and Bean-level hatred of foxes. Seeing them creeping about was always a call to action for everyone to run around the garden screaming. Even now

I think foxes instinctively see me as an enemy. They mark my doormat as their territory like the mafia. Late one night a fox stopped in front of my car in the road, made direct eye contact with me, and did a poo as menacingly as I've ever seen a creature do a poo, before casually trotting away. Damn and blast that lousy fox.

So why was this book delayed until now? For one of the same reasons this book came into being. Around 2008 my mother was diagnosed with breast cancer. Along with doing everything possible to treat it, she resolved to do everything she was expecting to have a lifetime to accomplish. She was trying to write at least four books at the same time, including a memoir and a murder mystery that didn't get past the first outline but involved a lot of dogwalking. The book you are reading is based on multiple talks and presentations she had made on henkeeping, and is based on a lot of her immediate experiences, and so was the quickest to finish.

By the end of 2011 it was all ready to go. And then in early 2012 she died. A silence entered our lives. A rocket that had propelled us forward for so many years suddenly ceased to be. The dog died not long after. The foxes grew more bold. It was too much. We gave the remaining hens to a farm, where I hope they lived the rest of their life in peace. We sold the house and went our separate ways. We tried to move on. I'm sorry to say we no longer have the space for chickens.

Nearly a decade later, a global pandemic spread across the world. We were confined to our homes. As a freelance filmmaker who could no longer go outside, I suddenly had a lot of spare time. I also discovered that my friend Alice Sage, an exceptionally clever and kind and driven person, had started her own publishing company. A thought stirred. I looked through hard drives and subfolders and eventually found something called KEEPING HENS_BOOK_FINAL.docx. It was funny, well written, informative and personal, everything that was so good about my mother's writing. I sent it to her and asked if she liked it. She did.

And here we are now. I will try not to interrupt the main book too much, but I will pop in with a few disclaimers. There is a lot of legal discussion now regarding the correct feeding and disposal of hens, and I will include details of this later in the book.

I have tried to track down which bowl was broken and who broke it. We are all very clumsy and our house had a lot of knickknacks. It was likely the bowl of an antique inkwell, and it was probably my brother Charlie, a big tough rugby player who opened a bedroom door too fast and knocked it off a sidetable. The thought of my mother being so apoplectic with rage that she angry-typed a mention of this into a book about henkeeping tickles me. It's also one of the only times Charlie is mentioned in the book, so let's clear that up now. He would like me to mention that he spent the most time cleaning up after the chickens, and cleaning the henshed, and claims to be the only one that could pick up the cockerel without fear. I would like to mention that he's an excellent human being who is bigger than me both in his superhero frame and his maturity. He currently lives in Somerset and has adopted some pheasants that fell into his back garden, which is kind of like keeping hens.

I have spoken less about my dad than my mum, but

Andrew Tribble is still alive and I also love him very much. He is currently quarantining with Henry in a small house in Aldershot, and is happily doing what he enjoys most in the world: slowly taking a motorbike apart, adjusting one piece, then slowly putting it back together again.

I picked up on reference to audiobooks in the text, and contacted the lovely and helpful Edward Peppitt, who originally recorded it. I was so incredibly happy when he found it, actually read by my parents and including audio from our real garden and our real animals. My father sounds like a jolly inventor even as he's talking about rat poison. My mother speaks with a beautiful regal radio announcer voice, which totally belies the fact that she grew up on a Peckham housing estate – her mother worked for the BBC and learned to speak in received pronunciation, and mum received her received pronunciation. The level of emotion in my mother's voice honestly took me by surprise. As I was doing the washing up at two in the morning, I listened to her talk about cradling a single hen in her arms as the others gathered around her. I stopped. I heard her voice break. She was starting to cry. I cried too. I thought of her by the old fireplace singing softly to a cuddled up hen. And then she started talking about hen bras as if nothing had happened. She loved those chickens so much.

If you're listening to this as an audiobook, hello, I hope my voice isn't annoying, and I hope you enjoy the rest. If you have this as as a book book, I hope you enjoy reading it as much as I did. If you are interested in keeping hens I hope this gives you lots of useful advice. The British Hen Welfare Trust tells me that right now there are thousands of chickens looking to be homed, and I hope this inspires you to consider it. I would be delighted if anyone uses my father's chicken shed as inspiration for their own. I hope you make peace with a cockerel. I hope you nurse a hen back to life on your lap. And I hope this leads to gardens full of happy, cared for, cheerily clucking chickens whenever the sun is up.

May 2020

Preamble
by Jane Furnival

You may be considering keeping hens. Maybe you bought some on the spur of the moment at a country show or garden centre, and don't know what to do next (it happens). Hens are a highly entertaining subject, even if you can't have them – yet.

This book is based on hen-keeping talks which I've given over the nine years. To tell the truth, I thought this stuff was all quite serious and straightforward, until I started getting requests for 'that funny talk you give on hen keeping'. Then, when fifty-five older people from Purley Literary Society turned out in the snow to hear this funny talk, I thought it was time to write it down as a book.

I am not a qualified poultry keeper, just a normal wife and mother of three, but I'm a pet-lover and don't kill and cook my pets, so this is not for the serious small holder who wants to eat his hens. In fact, we once tried to cremate a favourite hen in a ceremony but were put off by the smell of roast chicken. Disaster.

My own flock has thirteen birds, but has been as low as three – the minimum number of hens to keep as they are social birds.

Its leader is Winchester, a vainglorious and noisy rescue cockerel. As the cockerel is a large bird, Boo our Bouvier des Flandres dog who herds the hens, does not normally

bully him. My husband would like Winchester bullied by the dog, as Winchester intimidates HIM. I just want everyone to be happy... as long as the animals come first.

In Joyous Memory of Jacquetta Clark, who introduced me to hens in her garden in Brighton.

Thank you to my literary agent, Jacqueline Burns at Free Agent, for suggesting recording this book and lots of encouragement and help.

Thanks to Edward Peppitt for recording the audiobook and for all his help and advice in putting this book together.

Thank you to Peter Tarry for the portrait of me on the back cover originally taken for The Sunday Times.

Thanks to Jane Rodgers for kind permission to use a photo of her daughter Hannah Rodgers, to Anna Hamilton for kind permission to use a photo of her daughter Julia Hamilton, and to Melanie Lee for kind permission to use photos of her chicks hatched from my eggs.

With tremendous thanks to Andy Tribble for building the henhouse.

Awed thanks to Will Tribble, Charlie Tribble, Sonia Cooke and Henry Tribble, and Monica Gibson for feeding the hens while I was having chemotherapy for breast cancer and could not do it; and to Alexandro Aristizabal Molina, Lyn Campbell, Julie Mattin and Barbara Baxter for cleaning the henhouse for me.

Thank you to John Harrison at allotment.org.uk for permission to quote him.

I should be nowhere without Boo, the Bouvier des Flandres, Hen Herder and Fearless Finder and Feed-Sampler of Hens, and Fox-Killer, Wrist-Watch-Tapper to me to remind me of the correct times to take them out and put them away? My golden boy. No finer dog exists. Especially memorable was the day the hens lined up in a row along his back to keep warm! Wish I'd taken a photo...

Any mistakes are my own fault.

Keeping Hens

Before beginning to keep hens, to save heartache, check with your local council that they have no problem with your keeping hens in your garden. Stress that you are not going into a huge business and be flexible about a cockerel in your flock.

The biggest obstacle to keeping hens, in people's heads, is the idea that they need a cockerel. You don't need a cockerel to have happy, well-laying hens.

There may be a noise problem if you have a cockerel. I find that giving out fresh eggs to my neighbours keeps everyone happy. We're lucky, but you must respect your neighbours' sleep needs. If you're desperate to keep hens, move home. I would do this, rather than stop keeping hens.

I've had long times without cockerels in the flock and I had both our cockerels by accident.

If you don't want to talk cockerels but just hens, skip the next bit; but if you've got time for a meander through life and a bit of a laugh, read on.

The first cockerel, Mr Clumsy, was sold to me as a female. It was a genuine mistake, and happens occasionally, then to my fury, he started to crow.

He lived a lovely life for several years – till the fox heard him free-ranging in a far corner of the garden, and came for him and he sacrificed his life standing crowing a warning for his hens to run for cover, dear boy.

I should rather my hens enjoy free-ranging, and risk the odd sniper attack by a fox, than lock them in a run where they peck up all the grass, but other people have a different approach – and that is fair enough. They will be safe and hens do make the best of what they are given.

The second cockerel, Winchester, was thrown out of a moving car in a bedraggled state, presumably for losing a cock fight. As it happened, he was picked up and taken to my vet, Frean and Smyth of North Cheam, where kindly

Russell nursed him back to health.

This happened just after I suffered one of only three big murderous visits I've ever had from Mr Psycho the Fox, who struck when I was delayed by traffic in the Dartford Tunnel, stupidly left the entire henhouse open that day to get some sunshine in it, and could not phone anyone to put the hens away, as the mobile phone did not work in the tunnel.

If I had just left the pophole open, which is the small hole by which the hens can get in and out of the henhouse, I am sure the hens would not have been so decimated.

A few minutes is all a fox needs. Within that hour of traffic delay, I lost all my lovely girls, their headless bodies dumped back on my vegetable patch the next day. I especially grieved over Grayling, the granny of the flock, one of my very first girls.

You do have to get used to your birds dying in various ways, when you keep hens.

My son Will searched for survivors and at 1am, he found a hen dumped deep in some bracken, still alive but with her throat cut. After a fruitless night trying to nurse her, I took her into the vet to be put down – no, I will NOT wring her throat. How barbarous. Needs skill and not funny. So Russell the vet knew that I had an empty henhouse.

Two days later, I had cleaned out the henhouse and left it in respectful memory of the dead hens. Then the phone rang. Could I hear crowing on the other end?

'Jane, it's like a farmyard in here,' said Russell. 'I've got this cockerel, nursed him back to health. He's a fine bird, but no one wants him – so I'm about to put him down'.

'Russell, I don't WANT a cockerel,' I wailed. 'Tribs, my husband, will go mad'.

'Fine,' said Russell. 'If you feel like that, I'll just put the bird down'.

'NO! I'm in the car!'

When I collected him, he WAS a fine bird too. Bigger than an oven-ready turkey.

'What's his name?' I asked.

'Dave'. 'Dave?' I echoed coldly. 'You can't call a dignified and majestic creature like this, Dave'.

I addressed the bird in person.

'What's your name?'

'In… es… her,' he said with great concentration.

'His name,' I said with dignity, 'is Winchester. Can't you hear?'

'Do you oil him?' asked my friend, Diana Hallstrom when admiring him later.

Good question. A cockerel gleams gorgeously, except during the autumn, when they lose their fine feathers and look a bit bedraggled, like a woman who has emerged with her hat knocked sideways from the last day of Harrods' sale. But do I look like the kind of person with time to oil my cockerel? What kind of oil would one use?

Winchester settled in quickly – once I had hastily bought eight well brought-up young hens for him. What an escape, a rags to riches story for him. All he had to do was rape these girls in quick succession and he soon had a submissive harem.

It was an upsetting sight for my friend Anna Hamilton who is dickering about whether to keep hens or not.

'Jane, stop him doing that! She doesn't like it! Naughty boy!' she cried as she leant over the henhouse door watching the process.

It is over very quickly, but that is how a cockerel keeps control of his flock. It is nature. Don't diss it. The hen merely shakes her feathers and forgets.

In his opposition to the cockerel, my husband Tribs has reason on his side. Cockerels are life's passengers for hen keepers. They don't lay eggs, but you have to feed them. The hens claim not to mind if they don't have them, like old ladies at a care home.

However, actually, hens are rather keen on a cockerel. He will conduct them to the best places and show them titbits of food in a courtly way. He will keep a lookout. They compete for the position of favourite wife, which they enjoy, as it involves sleeping next to him on the perch.

A cockerel is highly entertaining for me, as he lies sybaritically in the sun with the wives giving him a dustbath. Here he is being tended to by two of his ladies.

Will says,

❝ *I didn't know what the word 'sybaritically' meant.* ❞
According to dictionary.com it is 'characterized by or loving luxury or sensuous pleasure'.

But a cockerel may not be so entertaining for men. I am convinced that cockerels have a sophisticated sense of smell, and they find men a threat and will fly at their chests, even in a half-cocked way.

A broom will ward them off. My son Will thinks like a bird, opens an umbrella and pretends to have better plumage, which scares them. Cockerels are real wusses. Occasionally I seize mine and just carry him round with me for a bit, to show him who is boss. He is slightly discomforted by this.

The more you feed an animal, the more he is your friend.

Top Tip If you do want a cockerel, you can pick one up for free or very cheap as lots of people don't want them. Look on the internet.

Hens are farm animals – they are only semi-domestic and really love to free range. They seem to understand about our boundaries pretty automatically – or if they fly off to visit friends in the next garden or stay out clubbing or something, you can clip their wings. More on this later.

Once, they saw a fox emerging from the roughs of the

garden and a passer-by knocked on the door to say they were walking down into my village high street as a flock. 'Just going to the charity shop for an egg cosy,' I quipped.

'Come quickly! Winchester has walked out of the front gates,' said my husband Tribble, appearing in the kitchen.

I shot out of the gates in time to see a flashy tail feather on the other side of the busy road. Winchester, our rescue cockerel, had walked out very determinedly, after a disagreement with our dog, Boo, whose task is to look after our hens.

I dashed after him. He was only stalking along the road to the Churchyard – where there are lots of foxes – and I was running, but I could not catch up with him.

Suddenly, with great purpose, he turned right – into the Village Library. The glass automatic doors opened for him and he stepped in. Then they closed on my nose, so I could only watch him stalking along the carpet towards the librarians' desk.

'Please don't do your business on the carpet. I'll get a letter from the Council,' I silently prayed. The librarians didn't look up, although one woman reader was silently

touching their shoulders and pointing. The glass doors opened for me and I found Winchester round the corner from the library desk, perusing the children's books.

Presumably he was seeking a book on how to deal with an irritating dog. Then he browsed the outsize children's books. I could not corner him, as he fluttered away from me nervously and he was clearly interested in the books.

Then he did what young men are tempted to do. He flew up and tried to photocopy his tail feathers. I seized him from the top of the photocopier and he let out a squawk.

When hens or cockerels feel cornered, one immediate

defence reaction is to loose their feathers, so their attacker ends up with a mouthful of fluff.

Keen to avoid this embarrassment, I tucked him firmly under one arm so that he felt safe, stroked his sparkling feathers, and softly crooned his favourite tune, 'Speed bonny boat, like a bird on the wing'.

Winchester always inclines his head and goes silent when I sing to him, listening judiciously. *Then* the librarians looked up, alright.

'Don't mind me,' I said brightly. 'It's just me and my cockerel'. As we vamoosed through the magic opening doors, I could see their jaws dropping at their table.

Back at the henhouse, Winchester was greeted by his wives with a chorus of clucks and reproaches along the lines of, 'What kind of time do you call THIS?'.

Some communication transcends species and hens are very good at making themselves understood.

Top Tip   I believe you can play Radio 2 for a calming, beneficial effect to a henhouse, but not all night – even hens need a bit of hush.

Eggs

Why keep hens? Producing your own fresh eggs for the family will be top of your list, I think. Are your own eggs worth going to the trouble of getting hens for, or should you simply buy good shop eggs?

'You couldn't be more pleased with that egg than if it had been your own,' said my husband when I jubilantly carried the first egg ever laid by one of my hens, back to the kitchen. The thrill!

Here's a test you can try at home. Compare the yolks of eggs that are battery-laid with

a) barn-laid eggs

b) free-range eggs

c) organic free-range

d) and if you can get them, your own or someone else's home-laid eggs.

Once you have your own eggs, crack all types in a plain white bowl separately in good light. Examine the quality of the whites for firmness, and whether the yolk is yellow or not. The yellower, the better.

Organic free-range hens have been kept to the highest agricultural animal kindness standards. That is the only type to buy, if you're not yet ready to keep hens at home. Sorry to ask you to shell out, but it's a vote for humane animal treatment and the only one that supermarket buyers – who are our farmers' masters – understand. The Universe will repay you for your kindness, another way.

When you have seen rescue hens reduced by a human being to nakedness and misery, as I have, you will understand why this cruelty must stop and for the sake of the farmers' souls too.

Health Issues

It is said that scientific evidence has shown that free-range eggs contain higher amounts of folic acid, vitamin B12 and vitamin A, than battery hens' eggs. Your own hens, happy and fed all sorts of nutritious scraps, may improve on that again.

Many people are concerned about the dangers of hormone and antibiotic residues in shop-bought eggs. 'Free range' is better than caged, but it doesn't mean organic.

John Harrison, writing a Poultry Page on the website www.chickens.allotment-garden.org, claims: 'In 1999 0.5% of eggs contained residues of Dimetridazole (DMZ). This is not even licensed for laying hens and is suspected of causing cancer and birth defects. At the same time, one egg in every dozen contained levels of Lasalocid in excess of 100 micrograms per kg. Again, this is not a licensed product for laying hens. The government watch over our food safety by testing one in every 18 million eggs consumed, which is hardly reassuring'.

Your own hens, under your control, may give you the best-tasting eggs and certainly the freshest. The taste quality of the egg yolk depends, I think, on giving the hen the ability to get out and eat grass. Greens, even if the edge of a cauliflower or the leaves of broccoli, are important for egg development.

 Some people suggest hanging a cabbage on a hook and think the hens enjoy jumping up to it. Just take the leaves and put them in the henhouse for them to shred. They are hens, not performing monkeys.

Will says,

❝ *Many henkeepers think mum was being a bit harsh* **❞**
about this! Gaynor Davies of the British Hen Welfare
Trust felt that, 'Actually, hanging some cabbage up
whole gives hens more enrichment and as it takes
them longer to eat it is a boredom buster'. Let your
hens eat cabbage in moderation (it can cause loose
droppings) however they choose to enjoy it.

The other factors that go towards good egg production are
access to light, preferably sunlight, and warmth. I don't
like some modern expensive designer hen houses which
shut them up in the dark. I shouldn't like to be shut in the
dark in a barrel. Common sense, isn't it?

A lovely yellow yolk makes gorgeous cakes but you will be
spoilt. I never eat hotel eggs now. If you want huge juicy
yellow yolks, give them sunflower seeds regularly – as a
treat, not a staple diet though.

I find that the best-treated hens produce a better-quality
egg. It is firmer, less runny and the yolk is yellower. In
the first year, I notice that our hens often lay huge double-
yolkers, which are nice gifts for children.

Feeding

An important note from Will.

" There is a big legal disclaimer I need to put here: the **"**
UK Department for Environment, Food and Rural
affairs has declared it illegal to feed chickens kitchen
scraps from anything other than an entirely vegan
kitchen, and illegal to feed chickens leftovers from
restaurants. It's a law attempting to prevent any
kind of health contamination for both the chickens
and people eating the eggs. I am writing this because
my mother will now go on to recommend feeding
hens kitchen and restaurant scraps, and in fact you
shouldn't do this.

In my mother's historical defense many henkeepers feed
hen scraps. I spoke to many hen experts who thought this
law was extreme, or too wide a blanket ruling. Some saw
it as something that makes sense for a professional chicken
farmer feeding hundreds of chickens and shipping eggs to
supermarkets everywhere, but shouldn't be applied to a
home keeper with a few chickens in their back garden and

a fridge shelf's worth of fresh eggs. Nevertheless, right now that is the official ruling. If your kitchen is entirely vegan, you're good to go. If not, it is illegal to do it. And never feed chickens scraps from a restaurant.

I would prefer not to alter my mum's writing, and she can no longer edit it herself, so I'm going to leave it as it is but remind you it is for anecdotal purposes only.

As far as I'm aware both the hens and anyone who ate the eggs were fine, thankfully. The only person who complained was the dog, who thought he should have first grabs at everything. Boo might turn his nose up at food put in his bowl inside, but if it was given to the chickens he'd run outside and make a big show of sniffing through it for anything he felt he deserved, making the most amazing protesting grumbling sound I've ever heard an animal make. It got to the point that he started trying to eat chicken food just as a show of authority. To avoid this we started varying putting the food outside the house and putting the food inside the chicken shed, but Boo caught on to this. Then as soon as the chickens were out of the shed in the morning he'd sit inside it, peeping his big furry dog head out of the little chicken sized door moodily, to make sure we didn't pull a fast one.

Oh and the other person that complained about our chicken feeding arrangement was a poor spider I once rescued from the kitchen and safely put outside the door, only for the brood of chickens to dash over and assume that anything put outside was probably food for them. I'm so sorry spider, I never meant for your rescue to end in disaster.

Anyway I'm getting off on a tangent. Back to my mum's writing.

You cannot just feed hens on scraps. The way the hens' biorhythm is constructed, you need to feed a solid commercial hen food in the morning, to enable the hen's body to make eggs during the day for the next day. Inside the hen is a conveyor belt of eggs, waiting to come out.

A hen needs a basic layer's mash in the morning when she is forming her eggs in her body. This is actually not a mash; you serve it dry. Hens get kerfuffled by wet things and have to keep wiping their beaks! You feel like handing them a napkin sometimes; they are quite keen on having clean beaks.

You are not wise to scrimp on buying a basic commercial hen food and try to mix your own, unless you go into it thoroughly – and that may cost more than just buying commercial hen food, which is cheap. Watch out for delivery charges though – they can slam up the cost considerably. I use Wiggly Wigglers because they deliver free for smallish orders.

Your hens may not lay well if they aren't getting what they need. I know that in the War, they boiled up potato peelings but I've wasted the fuel doing this, and I've never found a hen to touch them. If yours will do it, great, what's your secret?

We give our hens breakfast of Wiggly Wigglers Henfood with Bokashi, an additive which is supposed to make their droppings smell nicer. Can't say I've noticed.

But rather than keep our leftovers hanging around, we also add in last night's leftovers on top of their breakfast. Banana skins. Chips. The milk from the children's cereals. Hens, usefully, will eat sour milk.

You are meant to give them treats and extras in the afternoon, and during the day, more leftovers appear in my little kitchen bowl by the washing-up sink ready for this. It's a bit like giving children porridge or muesli for breakfast to give them sustainable energy, and a cake for tea, for extra pleasure and energy. But to my husband's fury, I tend to put things outside the kitchen as I find them during the day as I quite like to see them rushing to see what I've found them.

This can lead to hen invasions, as certain of the cheekier breeds of hen parade in through an open kitchen door and help themselves to dog food, driving one's husband wild as he has just cleaned the kitchen floor beautifully and they make a mess. People on the phone to us will find themselves interrupted by the cry, 'OUT! No hens in the kitchen!'

Here, at The Old Rectory, Cheam Village, we run quite an ecological garden. That makes it economical too. The hens eat our scraps; their droppings make good compost for our vegetables, which they eat the scraps of – and round we go again.

All our food scraps apart from anything highly spiced or oniony or chicken, go to the hens. The reason we don't feed them anything very strongly flavoured, is that the flavour comes through in the eggs. I expect sophisticated chefs to try feeding hens different flavours of food to get these nuances in their dishes.

When I am in a restaurant or café, I ask for my scraps and my friends' to be parcelled up as a hen bag. I had to be restrained at a café, from asking people if they had finished with that bit of salad... I was shameless, but my friends couldn't take the embarrassment.

I believe it is illegal to give hens meat but they do like anything accidentally dropped like bacon fat. Cat food that our picky cat has simply sucked the sauce from, is a great favourite, should they come across it. Dry pet food may stick in their gullets.

Will says,

" *Again, these rules have changed, and you can't legally* **"** *feed chickens restaurant food, food from your non-vegan kitchen, or bacon fat.*

They can eat too much of anything, even grass, and get a ghastly thing called 'crop-bound' where their throat looks swollen and sore. Not comfy, but really really hard to get rid of.

Storage is important. Don't put food where rodents can

nibble through it – and use a metal container, not plastic bins for that reason. I buy metal bins in two sizes. The smaller size fits oyster shell for the hens; the larger, the hen food.

Will says,

44 *You can buy similar metal kitchen bins in different* **77** *sizes from Habitat or Garden Trading.*

Also if you think any food has gone off – it smells a bit stale – you have to compost it and give them the right stuff. Hens are not dustbins.

What can you feed? Cheese, baked beans – and they think spaghetti is worms. The tops of carrots, cooked or not, chips (not too salty), cake – be careful of anything with alcohol – breakfast cereal...

If you should run dry of food while awaiting a delivery, I give mine muesli or breakfast cereal, though they may stop laying temporarily. It is very important indeed to give them salad and greens.

If there is strife in the henhouse, I find giving them lots of little pots of food, like old yoghurt pots, distracts each hen and stops them attacking each other which can be upsetting.

Another vital thing to give them is grit. This you have to buy, though garden hens will always use what they want from natural pecking too.

I buy oyster shell from Wiggly Wigglers and pour it on top of their food in the mornings; it is finer and easier for hens to digest than heavier grit.

Apparently you should also give your hens granite or flint chips that are insoluble to help them break down food in the gizzard – hens famously don't have teeth. A hen can consume 5 grams a week, left in a bowl in the henhouse rather than mixed with food.

You can get food supplements with grit mixed in already. One such supplement contains oystershell, redstone, limestone, korrel, seaweed, stomach stones, minerals, clay, high protein pellets, millet, biscuit flour, toasted soya beans, beer yeast charcoal, egg food, vegetable balls, herbal extracts, wild seeds, peeled oats, aniseed.

Will says,

❝ *The British Hen Welfare Trust recommends Gastro* **❞** *Grit from the Little Feed Company.*

Finally you must give them water. I change mine at least twice a day and if I find it dirty, I will refill it.

Worming

Hens need worming approximately according to the instructions on your wormer tablets or powder. Many professional wormers expect you to administer a tiny amount of powder each day for a week to a squawking bird – and also you cannot use the eggs for a specified time. Try Net Tex Gut, a herbal wormer which you spray in the drinking water and says you can still eat the eggs laid during the worming period.

Will says,

" *Gaynor Davies of the BHWT adds: 'There is only* " *one licensed wormer Flubenvet. Nothing else works to be honest. It is safe to eat the eggs while under treatment'. Flubenvet can be obtained from vets and farm shops that employ an SQP after completing a questionnaire.*

Odd-looking eggs

When a hen is starting out laying as a teenager, so to speak, she will lay tiny eggs without yolks – which elderly hens also lay. Some hens who are having a bit of a body clear-out or not getting enough to eat, lay eggs without shells. Yuck. Clear these sacs away fast or they will nibble them. It may be a one-off but it is best to increase the range of grit on offer as this affects shell quality.

The spectre of 'own-egg eating' is something that preoccupies professional hen keepers. They say that once a hen starts eating her eggs, the whole henhouse goes into one. I can't say I've often seen hens eating their own eggs. I've had crows fly behind me as I am collecting eggs, and steal a cache of eggs.

I reward Boo, my dog, for rounding them up, with a fresh egg straight in his mouth and if left, he will sneak into the henhouse and peruse the laying bays to choose his favourite egg – never anything from the fridge, please, only freshly-laid and preferably warm! That's fine, it's Boo's dibs for being so helpful.

We do occasionally put down boiled egg and scrambled egg leftovers which they enjoy and don't seem to connect with raw egg eating. It's probably some forbidden thing though.

The colour of the eggshell depends on the breed of hen, though a young hen always lays darker brown eggs. It's best to ask the breeder or seller about the colour of eggs a hen is likely to lay – they go from white to cream to blue to green to brown. Egg colour does not match the colour of their feathers.

Any hen with Legbar in its name will lay blue eggs, I believe. Legbars of Broadway are my hen gurus and I defer to what they say. Any serious problems with keeping hens, call them up and ask them.

Many hen-keepers complain that their hens don't lay in the winter; I once heard of Hugh Fearnley-Whittingstall saying this, to my surprise. That is partly due to natural factors – none of us are that productive during winter – and partly because the hens are cold. They adore warmth. Not the way that battery hen-keepers give them warmth; so much warmth that their combs tend to flop down and be pale.

I found this from my battery rescue girls when they first arrive. Unless it's part of the breed's natural looks, a healthy adult hen has a pink upright comb on her head like a comb or tiara. Growing hens who are not ready to lay, have a tiny one!

I put a greenhouse fan heater in my henhouse in winter and my hens loll in front of it and lay consistently well, throughout the most snowy times. They complain and look meaningfully at the heater if it is not switched on.

Because hens are susceptible to wet, and I don't want colds to sweep through the henhouse, I keep the heater on extravagantly if it is raining heavily, and keep them in.

Top Tip There is NOTHING so curious as a hen. Mount the greenhouse heater on a piece of board and put it high up away from harm, or they will peck it. Also don't turn it up so high that you risk fire from the surrounding straw etc.

I should not use any kind of fumey heater like a paraffin one; hens are very susceptible to breathing trouble and ventilation is vital; also you don't want to get fumey eggs. I buy a 2kw greenhouse heater on Amazon.co.uk such as a Gardman, but whichever you buy, they all seem to break down quite soon, an annoyance and expense.

Will says,

❝ My mother's opinions on heating are contro- ❞ versial. Henkeeping expert Francine Raymond told me that she would never 'lead people to believe that a henhouse needs any sort of central heating'.

Apart from eggs, why keep hens?

They are great fun, and make me feel happy and relaxed. If, for instance, your son has just accidentally broken your favourite and irreplaceable vintage bowl, a period of quiet reflection in the henhouse does wonders for your morale.

They use us like TV, standing on the window sills and looking in, and that is entertaining.

Also, they interact with other birds and that is fun to observe. They are very keen on protecting their interests and we often laugh at the antics of a couple of visiting crows who edge up behind them, hoping to steal bits of food, like playing the old game, What's the Time, Mr Wolf? – only to be found out and chased off, crossly. Hens are large birds.

Children adore them and so do guests. They are very, very soft and cuddly, not scratchy at all as I'd imagined, but they can struggle and fuss when you first pick them up. You have to nestle them against your chest and then they calm down and go to sleep. They LOVE being fussed and stroked on the whole. They are great family pets.

It is lovely to take six eggs rather than a bottle of wine, when you go out visiting – that is a prize above price! Especially if you explain that they have only just been laid.

Or if you have visiting children, there is something magical in allowing them to find eggs in the henhouse, wrapping them up, maybe in bubble wrap or kitchen roll, with their name on the front.

 Let your friends choose some hens' names or even name them after your friends. It pleases them, even if your hens secretly have multiple names or you forget.

Hens are moving garden ornaments. They look gorgeous with their fluffy duvet-like coats, moving across the lawn in the morning or the evening, as a flock.

They are friendly on the whole. When gardening, you hear a little voice saying, 'How are you?' next to your ear. When you are digging, they will congregate around you.

Talking of excavations, they are definitely not stupid, nor

do they have short memories. Hens are equipped with the knowledge that hens need, to be effective hens, not jump through hoops because humans have decided that's a test of intelligence. They can learn and remember for years – for instance, an attraction to workmen in yellow coats. My previous flock learned that a bright yellow coat meant DIGGING and paraded across a busy road to raid a hole for worms and insects.

Beware putting your young seedlings into pots and leaving the greenhouse door open. I've been trying to plant up an old sink with herbs for years and they simply scratch anything out and sit in it, giving their best friends long dust baths and gossiping and singing in the sun. Bliss.

When they are lying down at their ease with their wings extended, having these dust baths, they can look as if they are wounded at first sight, but don't panic unless you see distress signs such as an open beak, which means panting and needing water.

Even if there is nothing edible for them, they are more than capable of scratching until the entire seedlings are uprooted. I wish I knew the rhyme or reason of why they do this or what plants they choose!

They adore dahlias, consume tomatoes and courgette plants, but once a plant is well-established, apart from these, they will leave it alone. They don't like carrots, garlic, potatoes, parsnips or peas in my experience.

The cockerel and the sunflower

My son Henry, brought home a sunflower seed. This proved part of a school competition to grow the best sunflower, so we planted it, fed it and put it on the kitchen windowsill to grow.

At that time we had a cockerel called Mr Clumsy. He was entirely brought up by me, from a youngster, and had no experience of sunflowers. He would perch on a garden table outside the kitchen windowsill and look in.

That sunflower grew rather nicely.

Then one day, I walked into the kitchen and realized it was gone. All of it. Not even the root. One slight footmark of dust on the sink betrayed who had eaten it. Mr Clumsy could obviously SMELL something good, even if he didn't recognize the sunflower among the various other plants on the windowsill, none of which had been touched.

Top Tip Protect your vegetable garden or cherished flowers with netting. Of course, take the netting down when you want your garden denuded of slugs and other pests, as hens are PERFECT for this. I rarely see a slug in our garden.

Then they manure your vegetable patch for free. If you don't keep your own hens, you can always borrow someone else's for a few hours.

How many hens?

Three hens will produce enough eggs for the average family of four. Hens and cockerels are flock birds and it is cruel to keep just one, though there are always exceptions. A friend of mine has an old hen she keeps in a chair in her living room like a granny, and they are very happy watching telly together.

Three hens is really the minimum you should keep for their psychological health, and is a number that provides for their social needs. There will be a leader, a follower and the hen-pecked underling. This one will feed last and sleep at a distance from the other two, who will bully her a bit. That is how hens like it, I'm afraid, but like schoolgirls, their system does allow for a certain degree of social mobility.

I once had a hen called Madame Pompom with a slightly mad 'hat' of fluffy feathers on her head. She was ostracized by the others, despite her beauty, but didn't seem to mind too much. I got upset and my husband and son suggested buying a second one of her breed – goodness knows which, I'm not a great breed aficionado – to keep her company. She merely bullied the newcomer. I realized that I had to keep out of matters.

You often find that a weak hen will pal up with the strongest hen too. They obviously chat to each other.

Words like 'cock of the walk' and 'hen-pecked' come alive when you have hens. The outsider hen tends to get last go at the food, and sleep in the least nice positions, i.e. away from the cockerel if you have one. You will often find them pecking a little bit away from the rest of the flock. These social positions can change however. You just have to be as nice as possible to the outsider – I give mine little extra treats and she is treated with a bit more respect.

Very young hens are all the commercial hen-farmer will stock; they will slaughter them at anything from eight months, I understand. They are still good layers and nice hens, very grateful to be rescued – every animal has a personality – and we'll talk more about rescue hens later too.

A hen can live naturally until she is about eight, though I've heard of an arthritic hen who is eleven, and an old hen will still lay the odd egg.

Although you only get an egg a day from hens up to about the age of two, older hens are a valuable part of the flock. They represent grandmotherly stability and experience; they are the repository of good advice for the rest of the flock, showing them the best places to feed, advise on safety and protecting the hen-pecked. I should keep them on always despite the fact that the mean-spirited would call them passengers...or is that how they feel about their own grannies?

Feeding containers

Professional sellers of hens confide that they make a fortune on those galvanized traditional tin drinkers and feeders. These freeze in winter, and I find them fiddly.

You can buy expensive special poultry water drinkers, but I have found them unsatisfactory, though they look lovely. The tin ones also freeze in winter.

You are better with a plastic washing-up bowl weighed down with a clean brick – this is Terry Dunk of Withy Cottage Poultry's suggestion – so they don't kick it over. They kick everything over on the whole! Frustratingly, they will also drink from any old puddle!

Rats, mice, weevils, anything, can get into a stock of hen food left out for demand-feeding.

I like to feed my hens fresh every day. We use any food container that comes to hand before it is thrown away. You don't even have to wash them up, but don't give them anything that chicken or strongly flavoured curry has been in. Don't give them fibreglass takeaway containers, as they will peck and eat them.

Hens are not polite eaters. Once the hens make a mess of the containers, upending them, standing and messing in them etc, the entire container is then thrown away or if possible, composted. We particularly like cardboard pizza takeaway containers as you can feed out of both sides, but I use old chocolate boxes, even bits of kitchen towel.

When cleaning out the henhouse, hen droppings go on the compost heap to help grow vegetables and for the garden. Saving money on expensive fertilizers. People say, don't put hen mess straight on plants as it might burn them. I have never seen any sign of plant burn.

The economics of keeping hens

I worked out that, including food and bedding, but not including the cost of building our hen house, each hens' egg costs me around 3p.

The cheapest supermarket free range egg I have found, not organic free range which is code for 'best farming practice', is 14p from Asda; the price of organic free range eggs has actually increased to around 35p. You can buy an egg from a hen cramped in a cage with no feathers and fighting for her food, for around 8p.

Will says,

❝ At the date of writing (summer 2020), Asda ❞ continues to charge about 14p per free range egg on their website, but you can now get organic free range eggs from approximately 23p per egg, or even 20p for an organic egg if you go to Aldi. Officially 'battery hens' were banned in 2012, but they were replaced by terms like 'caged' hens instead (something I'll discuss later), and you can buy eggs laid by a caged hen for about 7p an egg.

Occasionally, if I have spare eggs, I sell them at 25p each, and that helps to pay for hen food. A small sign put up saying 'A few free range eggs from our garden for sale' normally brings enough business, but now I have ladies who say, 'Call me if you have 12 eggs and I'll drop in and collect them'.

By the way, when selling eggs, which is legal for small producers, you need to point out that they don't have the lion mark and tell buyers to eat within 14 days. That means that pregnant ladies, the vulnerable elderly and babies should avoid them in case they catch listeria.

 I ask guests to bring me their empty egg boxes, or if I have none to give eggs away in, will stuff eggs gently into biscuit boxes surrounded by kitchen roll.

I have had some labels printed by Vistaprint, who offer a very cheap service, saying 'Home made at The Old Rectory Cheam Village'. Sticking one of these on an egg box enables me to date the eggs and write a warning saying that they must be consumed within 14 days – of laying, not buying, so build in less time if the eggs you sell are a few days old.

Also tell people to wash their hands with soap and hot water after touching the shells – hens have dirty bottoms!

When gathering eggs, rinse them minimally as you are washing a protective layer off them. You can always wipe them gently down before putting them in the egg box for your friend. They can get awfully muddy at times.

Don't write on your eggs in any pen or pencil as they are porous. Write the date on the fridge door instead or label the fridge.

What else should you consider?

Your time. You or someone do need to be there to let them out, feed them and change their water, and lock them up and that includes holidays.

Unlike ducks, hens will make for their henhouse within plenty of time of sunset. There are timers you can get which close the henhouse pop-hole – a small inner door just big enough for hens but not foxes. I've given you the reference for someone who sells them at the end. Sometimes you can get back later than you thought, and the hens have safely put themselves onto their perches.

I don't make ghoulish jokes about killing or eating my pets as I like them alive. I can't stand the idea of wringing a hens's neck. Even with a machine you can get, it's not in my view a job for a pet-loving kind of person, things can go very wrong and the neck can untwizzle so you have tortured the poor bird. A friend of mine chopped off a hen's head in front of the others and they didn't lay for weeks and weeks. They went into shock. They are friendly and co-operative creatures who want to love us.

There is also the time you need to spend on cleaning out the henhouse. This starts out as a fascinating pleasure and quickly becomes an irritating chore.

Hen droppings are streaked with white, which is the urine part of it. I put down newspaper where they roost but my husband Tribs says it turns to papier mache. A wallpaper scraper is the best tool for cleaning up poo once it hardens. You can buy specialist hen cleaning products. I wipe down with a wash up sponge and throw it away.

Poultryshield is the standard cleaning liquid – which protects against the dreaded red mite, an itchy little parasite they occasionally get.

Stalosan F is another popular cleaner and must for the Poultry Keeper and claims to be effective against the spread of infection. including Salmonella, Coliform Bacteria, Streptococcus Uberis and Staphylococcus Aureus, fungi and parasites including roundworm.

If you want to stick to organic cleaners, Barrier VI says it is approved by the Ministry of Agriculture, Fisheries and Food and is even safe for day-old chicks.

Will says,

❝ *The Ministry of Agriculture, Fisheries and Food* **❞** *has been replaced by the Animal and Plant Health Agency (APHA).*

You can also dust down the laying areas with red mite powder against red mite, a tiny insect you sometimes find irritating your skin as a corps marches over your wrist. These are very irritating for the hens, and generally they should not be left in insanitary conditions, though we are not obsessive about cobwebs etc. Don't bother cleaning red spider mite with anything except the correct stuff. They have an invincible covering and normal cleaning stuff does not move them. They just hide.

Smaller henhouses need complete mucking out more frequently, but I do a huge clean with power jets, about once every three months and a clean-out and change their bedding when I think they need it. Buying one of those disposable cleaning outfits to wear when cleaning out, is the best idea.

Over winter, I leave them longer – it is ok to build up newspaper-based bedding on the floor, which they rearrange anyway to suit themselves, as long as it doesn't get whiffy. Add a new layer on top, and eventually it all gets trodden into a kind of cake and goes to the compost.

Do you want the trouble and mess? Hens tend to 'sour' the immediate grass around the henhouse. If you have a treasured lawn and you want to keep it like RHS Wisley, find another place to put them. But a charming henhouse is really an attention-grabbing thing too. They don't tend to eat things like roses.

Some people solve this problem with a movable ark or by moving the henhouse around, or a run they just keep for outdoors use.

Do you have enough space?

Even having just a balcony, though not ideal, is ok. I'd always say go for it rather than not, though I've encouraged someone to keep hens and she had to hand them over to me eventually, as she could not go up and down from her flat to her garden, and mess with them all the time.

Hens are tremendously adaptable – the Japanese keep them on tiny roofs. It's best if they can free range or wander when supervised, but not essential. Hen-breeder Philip Lee-Woolf of Clarence Court, whom I consulted, said you need at least one cubic foot per bird for sleeping sheds only, and at least 3 cubic feet per bird for a covered run with occasional letting out to the garden. You can get one and a half bantams in the space occupied by a hen.

If you have bigger flocks, you may get problems if there are outbreaks of avian flu, but a smallholder should not get hassled.

Dogs

Boo, our Bouvier des Flandres dog, is invaluable. His job is to make sure the hens are happy in the morning and to keep foxes at bay - which he does by jumping on their backs and killing them cleanly, then walking away. He will also sniff out any hen who has decided to hide.

Will says,

❝ *For safety's sake I should add that you legally should* ❞
not let your dogs kill foxes. If your dog accidentally kills a fox you will need to explain why it wasn't under proper control.

One young lady hen last year decided to nest down the side of the compost heap on a pile of fragrant rose petals. The hens suddenly decided that her unauthorized absences were intolerable. There was chaos in the henhouse that morning. Normally, like kids raising Cain at school, there will be a dreadful fuss - then when their human looks over, there will be silence.

But this cajoling and 'where's Doris?' clucking was going

on, so I gave Boo his favourite command: 'Find my hen'. He was off round the garden, then found the hiding place; stood there like Gromit, his eyebrows sad but firm, whilst I said, there's no hen here. He raised his paw, which is his sign that he politely disagrees with Darling Mummy.

I saw a flash of feathers and extracted a hot, cross hen from the side of the compost container. She had been sneaking away during the day and sitting on about eight fine eggs. Unfortunately at night she had gone in and I suspect rats had peed over them, so they had to go. Never ever be tempted to keep an egg that has been in the outside. It may be polluted by goodness knows what.

I restored her to the henhouse – where there was a lot of telling-off and wing-waggling – and took a planter in there, which I lined with rose petals.

I really don't mind a broody hen, but they won't get off their nests and they can kill themselves unless you give them a bit of separate food and water, under their beaks, which they grunt appreciation of. Eventually all my hens get bored and want to go out and see the sunshine, so I've never had naturally hatched chicks.

When a young dog, Boo has killed hens, under the impression that this was what I wanted him to do. It broke my heart. Then he went through a stage of just wanting to lick them, as if getting the flavour. It terrified one to death and it was only when he saw me crying over its body, that the penny dropped.

A lot of dogs will kill or bully hens; you need the dog really to come after the hens, not before, and you really need to train them carefully – especially terriers.

Top Tip Beware visiting dogs to your garden and always lock your hens away to avert tragedy.

I once found our cat poking the hens in the henhouse as it amused him to see them flutter. A sharp reprimand dented his dignity and he never appeared in there again.

Choosing your hens

So you have decided to keep hens. What kind of hens to choose?

This is the fun bit! Browse some illustrated books from

the library, but don't get too carried away as hen sellers usually don't have all the options. Different breeds have their own characteristics. Think what you want and what you need from a hen. A good layer will lay between 280 and 300 eggs in her first year but this will drop off in winter and as she gets older.

It is best not to buy hens from a country fair. Although tempting, you never know about infections etc and 'here today, gone tomorrow' breeders may use this method to get rid of their old or ailing hens or anything.

Do a bit of research locally. Make an appointment with a hen breeder, preferably one close to home as travelling is stressful for hens, although they go very quiet and you would not always know they are stressed. If you find they are opening their beaks, when travelling, it means they are under stress and thirsty. Water is absolutely crucial for hens – if they go without it for 12 hours, they can die.

I say 'breeder', but quite a few hen retailers these days receive the same deliveries in huge trucks from abroad, and 'bring up' junior hens. Some breeders will show you all the stages of junior hens from chicks. Some breeders will not sell outside of February to October, as they like to sell hens in the best condition and they moult in winter.

Take a box or basket to carry your hens away. Take a few more boxes, as you normally end up with more than you intended! Line them with newspaper and straw or hay. Not shredded paper which mats up in their claws and makes life hard for them. Dust free bedding is your best choice. Hens are sensitive to dust and ventilation and breathing. Some hen bedding is also super-crushed so that you don't need to keep buying tiny amounts but the downside is it is heavy and you have to store it.

 Whatever you do with hen bedding, they will rearrange it to suit themselves so don't bother styling it beautifully!

Different breeds have their own characteristics. Think what you want and what you need from a hen. I'm looking at my latest bill for hens, from Terry Dunk of Withy Cottage Poultry.

1 x   Withy Sussex, £15 (we called her Marilyn as she has a white skirt)

2 x   Withy Cuckoo, £30

1 x   Columbine – grey I think, £22

2 x   Withy Black Rocks – called after Barbara and Lyn who helped choose them - £30

1 x   Withy Ranga – Alice after Lyn's daughter- £13

1 x   Withy Blue - £15

Total £125

I once bought a black hen on the spur of the moment and called her Tribbly as her colour reminded me of my husband's hair before it went grey. She was a bad tempered bully of a hen. Even hens can go to the bad!

Pure breeds are pretty, but less hardy than hybrids. They can lay up to 100 fewer eggs a year, though the shells may be thicker and the quality finer for cuisine.

The ones with fluffy feet, I should beware of, as they tend to pick up foot infections, I understand. God knows what you do about those, but you want to keep their feet as clean as possible which is why I don't use shredded paper in the henhouse anymore – it's not warm, they don't like it, and it tends to catch in their feet and annoy them. You can mix a little bit of shredded with the normal bedding straw if you want to eke it out.

The reason hybrids were bred, was to combine the hardiness and high egg-laying characteristics of various pure breeds. For instance, Buff Orpingtons are the teddies of the hen world, large, friendly, fluffy, birds that look gorgeous, but are easily damaged by children apparently. Someone recently mentioned that she had never cuddled such a friendly cockerel.

I have chosen hens to be fast and noisy, on the whole, in case they are attacked by the fox. They make less entertaining pets. Now I just choose whatever I fancy. We always have a red hen called Chickeny named by my youngest boy Henry. They don't come by name if you call them, I find.

I let the family choose one each, chose two Sussex Lights

who are white with a necklace of black around them, because they are nice dependable characters, good layers, a nice all-round pretty hen, and they were pals and I didn't want them to be split up. Then I chose another one who the breeder said was a good layer – he only charged £13 for her rather than £15 and I think it was because she was a bit older; she is a good layer and a nice girl.

Don't bother with buying hatching eggs. It is expensive, difficult and you may get a lot of cockerels, who you must give away, as they will fight with each other.

Bantams. Not motorbikes! These are smaller and lighter than hens and great if you have a smaller garden. My father kept bantams in a barn and free range in his Welsh farmhouse. I don't know why he didn't just get hens, as the eggs are smaller and it made recipes difficult – you need two eggs for every hen's egg. They are also quite fast and flighty, and hard to catch. Friends of ours let their Bantams breed naturally in Norfolk and said it was like gang warfare with various young cockerels challenging other ones and groups fighting.

Don't bother with buying 'growers'. These are teenage hens who haven't started laying eggs yet and you can tell they're immature because the comb on top of their heads isn't fully grown. A hen starts laying at anything over 4 months. You want a 16 week old hen. Growers cost the same as layers or hens at 'pol' – point of lay. But growers need special food, growers' mash it's called, and you are just investing your money for nothing – when you can get layers or hens at point of lay for the same price and not have to buy the special growers' food.

You want your hens fully vaccinated and wormed, and ask them to clip their wings for you. A breeder will do this for you as a favour every season but it's not hard for you to do. Here is Terry Dunk demonstrating.

Easy and not worrying – it doesn't hurt them, just puzzles them! This stops them flying away. Clipping one wing means they can slightly fly, which is useful if they want to get out of the way quickly, but they go round in circles.

Your hen breeder will do it for you before you leave which is the easiest option. Of course, remember they moult in the autumn and you have to do it again next year!

Use sharp scissors and be quick. You cut into a section only of the wing, about 3 feathers from the front of one,

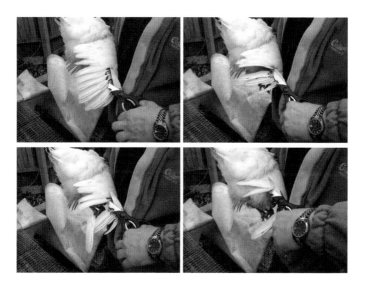

not the whole wing so that it does not show. Cut about three inches up in a section, and then leave about three feathers the other side of the wing.

Will says,

❝ Not all henkeepers clip wings – some find it ❞ unnecessary, especially for heavier breeds. Do what works for you and your hens.

The miracle of Melanie's chicks

Melanie Lee buys my eggs. She did not let on that she had bought an incubator and was trying to HATCH my eggs. The window of opportunity came at Christmas when our fridge packed up. The eggs were not chilled enough and stayed fertile.

Here are some photos of 'my' new chicks – the nearest thing I'll probably have to grandchildren! There are three hens and a cockerel so she has a ready family. What a complete thrill that turned out to be. But is it incest for them to live together? I asked my hen breeder Terry Dunk, who said, well, it's not ideal but just don't keep replicating the same stock.

They have grown very quickly and can now read the paper. Here is Julia Hamilton with a newborn. You can see how much equipment and time is needed to tend to the chicks, and Melanie paid around £80 for the incubator alone. Raising chicks is not a cheap option unless I suppose you rent the equipment to others.

Rescue hens

I have had rescue hens several times and it's a good thing to do – very emotionally rewarding to see a hen whose soul has been battered out of her, regain her personality and her feathers. They are the cheapest hens – they are free, but you're expected to make a donation of around £4 to £7.

Will says,

❝ *My mother refers to battery hens below. In 2012* ❞ *battery hens in barren cages were officially outlawed in the UK. Unfortunately other forms of hen caging are still common. 'Colony', 'enriched' or 'caged' hens are kept inside a cage designed for 80 hens at a time, with a nesting area and perches, where they will be kept laying eggs until the day they are slaughtered. 'Barn' hens are kept with many other hens inside a large barn-like space rather than a cage, but again they will never leave this barn until they die.*

For a naïve moment when I learnt that battery hens were banned I felt hopeful. But images of the of the new forms of caging don't look all that different to me. It brought back memories of the rescue hens we had, with bare scrawny necks, strange featherless patches, and bones showing through their wings. Mum would come home with them sometimes, and we would gradually acclimatize these poor nervous animals to standing in open spaces of grass and sunshine. I would like to live in a world without hens being caged.

Gaynor Davies from the BHWT says that, 'these days barn chickens are bred from very docile hybrids and are generally very gentle not aggressive. Goldlines are not common now. The hybrids more commonly used now are Hy-line, Lohmann Browns and Shavers'.

It's always a bit iffy introducing new hens especially battery hens, who are quite aggressive to other hens, as they are used to fighting for every bit of feed they've ever had. I called one of my ex-bats 'Aggie' as she was even aggressive to her fellow rescue hens. They will cower within the henhouse and sometimes don't ever go out much – that's what they are used to, you see. So if you're going to start with batteries, I suggest just getting batteries rather than mixing your flock.

You can't just go out and get a battery hen. You need to put your name down and wait sometimes quite a long time. So put your name down soonest. It is rare that you get a special breed – they are normally the red type known in the trade as Goldline, great reliable layers and hybrids. It is extremely rewarding to nurse them back to health after they arrive with a few feathers and not even knowing how to walk – the word 'cooped up' comes to life in front of you when you see a battery hen.

One thing to consider is heat. They have been given excessive heat, to make them lay faster, and if you simply remove that heat suddenly, they get colds and flu. I have considered adapting tea cosies but in fact, a knitted dog coat is best. Some never look like nice normal hens for a good year or more, if ever – we used to call ours Scruffy and Ugly I'm afraid, and explain their history to visitors, but they were nice girls and cheerful good layers.

I put my rescue hens – I last had three for Christmas – in our greenhouse and although they devastate it, it is a great refuge for them. I have found with battery hens, you sometimes lose several just from the stress of being moved and the huge change they undergo.

They come with the residue of various professional farming injections, and that's fine until it all wears off, then they are more prone to flu. I lost my last battery girls, one after another, two winters ago. I find if flu attacks, they die at the rate of one every two weeks.

Will says,

❝ *Gaynor Davies comments on this, 'We don't agree* **❞** *with knitted coats or tea cosies, they cause more problems than they solve - keeping a hen artificially warm may delay new feather growth, and they could get tangled in wet wool'. Gaynor also objected to the generalness of the term 'flu: 'Hens can get Avian Influenza but that is a notifiable disease akin to a pandemic and when we have it in the country hens are not allowed access to outside space. It kills the hen within days (hours sometimes). Hens can get Infectious Bronchitis but most are vaccinated against this (certainly commercial birds are). There isn't just straight 'flu' ...There is no value to a farmer to keep stock that is unhealthy, so the vast majority of birds will come out healthy if sometimes a little feather-bare'.*

Illness

Sometimes they pull through. But losing hens is something that is traumatizing to start with, then you get used to it.

Princess was my baby, my Buff Orpington, twice as large as the other hens. Though a junior, she quickly established that she was the boss. Then she started to mount them. She never laid an egg nor grew a proper red corona to show that she was a fully-fledged hen.

I called her breeder and said, she's mounting the others. 'I had one like that,' the lady said, 'then she died'.

Very helpful. Then she died.

It was only with the aid of a dove breeder, that I established the truth years later. Some hens are born DNA confused and change sex later. The stress of this on their bodies leads to an early death.

Princess stood outside the back door waiting for me to return home one afternoon, her retinue of hens around her. She looked pale – see the photo here. She jumped into my arms, nestled, sighed and died. I was so shocked. I sat down in a garden chair with the other hens standing in a semi circle around me, just looking and willing me to do something. Eventually they walked away and pecked in a meditative fashion.

A crop-bound hen

A battery hen I had, ate too much and became crop-bound. That meant that food just sat in a long pouch under her throat. It looked like a tumour but did not seem to affect her. I consulted everyone. I tried giving her yoghurt and goodness knows what, Alka Seltzer – funny seeing a hen burp – turning her upside down and massaging her all the time – she just leaned her long neck the other way and stared at me puzzled. It was like playing croquet with flamingoes in Alice in Wonderland.

So I found a hen specialist vet. These don't come cheap, I warn you, and need not be very good either to justify their massive cost. Mine constructed what he called 'a hen bra' of tape and bandage to try to keep her in, and gave me some medicine which did not work.

She pecked off the bra on the way home. The crop-bound condition stayed with her unaffected it seems, until she died of something else. She was a cheap rescue hen who cost me £180 at the vet.

What to put in your henhouse

Make sure that their perches are not plastic. They cannot grip them. An old broom handle is fine; give them lots of length. Move things around a bit every now and again. We use old pallets as ladders.

Let's think about the enemies of the hen, principally rats.

They say you are never more than six feet from a rat; we have 260 acres of Nonsuch Park on our doorstep and that contains a lot of rats who love hen food.

I once called a rat man to deal with the rats around the hens. He was frightened – of hens. He ran away.

Then there were the ferrets. Maybe some parts of the country have true rat-hunting ferrets. Not in Surrey. The ferrets came, in their pretty pink harnesses, sniffed around, and did nothing until their owner put them away in their cage. At that point, the rat *brushed past their cage* smoking a cigar. OK I lied about the cigar.

Rats
by Andy Tribble

A big problem. Rats can upset the economics of your hens if you have to pay an expensive contract to rat exterminators – these don't come free on the rates anymore.

For a brief time, I experimented with traps. The problem with those is that hens can step into them too. So I went to the trouble of only putting the traps out after the hens had been put away for the night and then gathering them up again in the morning.

There are two problems with this. Rat traps can be stepped in by your neighbour's cat and people have been sued for damaging their poor neighbours' pets. Secondly, leaving a trap out overnight for one night only is useless because rats are the *opposite characters* to hens. Hens investigate everything but rats are neo-phobic and don't like new things. A rat won't investigate a trap until it's been there for days or weeks.

There are box traps but for me, these have never worked. If you set about poisoning rats, you are often tempted to invest in a series of plastic boxes that you will then need to put the poison inside and leave for weeks until the rat gets used to the box.

I've found this a problem too – things ended up being piled on top of the boxes in the compost heap. I put a box in the compost heap but it was buried by rats six foot deep. If I laid a box under a pile of roof tiles I then had to disturb the tiles to refurbish the poison and that would alert the rats to the fact that I had been there!

At one stage, we contacted a member of a local gun club. He appeared to go about rat hunting in the same way as fishermen go about late night fishing. He proposed building a hide, dressed in camouflage, armed with night vision equipment and weapons designed to hit the rat but not powerful enough to go through the fence and not damage an innocent passer-by. Our so-called rat shooting expert gave lengthy descriptions of his silenced air guns for hunting rats in an urban setting.

We never saw him again. The rat hunter vanished.

Our rat problem became embarrassing because people started to notice the casual way that rats ran around in broad daylight and the rat would wait till Jane was in the shower and could see him sauntering into the henhouse to finish up the food. She would shoot downstairs wrapped in a towel uttering deep imprecations of rage, only to find the wretched little thing gone.

The solution was right under our noses. For a long time, I fought this bitter campaign against the tunneling habits of rats. They had opened up cracks in the concrete of the henhouse floor and I poured buckets of wet cement into these cavities, only to find them opened up again after a couple of days.

I suddenly realized that an unsightly open rat hole was best left open. Instead I poured poison down the open tunnel. The poison landed on a ledge six or eight inches down, out of the hens' reach so there was no risk of other animals picking them up. It was also familiar territory to a rat, unlike a plastic poison box.

All I did was stroll out to my rat tunnels every day and drop bright blue rat poison down it. After a couple of days of checking, the poison was no longer taken. After a week, there were no more rats.

A second advantage of rats dying underground is that there is no need to collect the bodies. I have to leave those holes open and if I suddenly see rats, I will repeat the poison. Easy. No need for traps, expensive contracts nor much fuss.

Will says,

66 *Francine Raymond advises, 'never bury rat poison -* **99** *a length of plastic drainpipe is better'.*

Rats are smart, and I'm sure some can be lovely, but wild ones that get in your living space can be terrifyingly huge. I am a Technically Londoner. Rats, like foxes, still occasionally sashay into my life, even without keeping hens. A rat got into a flat I lived in a few years ago, through a linked mix of a terrible landlord and a storm drain that was directly connected to a large hole in a kitchen cupboard. I heard rustling and dramatically threw the cupboard open, hoping to find a little mouse in the tiny humane trap I had set for it, only to see something big enough to reach from one shelf to the next. It had made a nest out of reusable plastic bags in the skirting board. We had to put down poison and seal up the cupboard with duct tape for a week. I'd be doing the washing up and would hear it hammering itself against the door, trying to get out. It never ate the poison but just got bored and left on its own. My other rat encounter was in Leicester Square – it was standing quite proudly on a wall, surrounded by tourists delightedly taking pictures of an official London rat. I think it was posing.

I don't have any tips or corrections to offer. I just wanted to tell my stories about rats.

Building a Henhouse
by Andy Tribble

Under some overgrown ivy in the corner of our garden, we found the base of a Victorian greenhouse that had a concrete floor and a brick surrounding wall about two feet high. Based on this, I did some drawings using library book reference. Jane's needs were that it was tall enough to stand in, making it easier to sweep out; another was that it would have a laying gallery along one side.

A laying gallery has a series of cubbyholes for the hens to go and sit in rather like a public loo and a little roof which hinges upwards so that you can reach in from the outside and collect the eggs without needing to go in through the main henhouse door.

THE BASIC IDEA

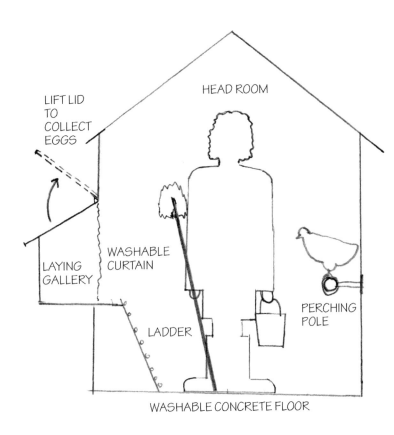

LIFT LID
TO
COLLECT
EGGS

HEAD ROOM

WASHABLE
CURTAIN

LAYING
GALLERY

LADDER

PERCHING
POLE

WASHABLE CONCRETE FLOOR

WHAT YOU NEED

HANDSAW
(OR A JIGSAW IF YOU'RE LAZY)

SAW HORSES

STANDARD HAMMER
& SMALL HAMMER

PENCIL

SPIRIT LEVEL

BLOCK
OF WOOD
&
SANDPAPER
(DON'T SPEND TOO
MUCH TIME WITH THIS)

THIS SHOULD DO EVERYTHING EXCEPT:

FOR WINDOWS & DOOR CATCHES
I DRILLED A FEW HOLES AND
PUT IN A FEW LONG SCREWS

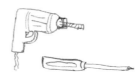

I MADE ONE OF THE RIGHT SIZE, LAID IT FLAT ON THE LAWN, HAMMERED IT, THEN MADE THE OTHER SEVEN JUST BY LAYING THEM ON TOP OF THE FIRST ONE AND MAKING THEM THE SAME.

THE ROOF BEAM FITS INTO THIS SLOT

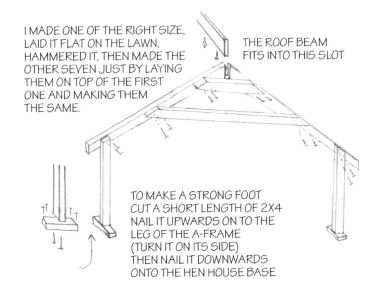

TO MAKE A STRONG FOOT CUT A SHORT LENGTH OF 2X4 NAIL IT UPWARDS ON TO THE LEG OF THE A-FRAME (TURN IT ON ITS SIDE) THEN NAIL IT DOWNWARDS ONTO THE HEN HOUSE BASE

MAKE A STRONG BASE BY LAYING TWO SETS OF 2X4 PLANKS, ONE ON TOP OF THE OTHER, THEN OVERLAP THE CORNERS LIKE THIS

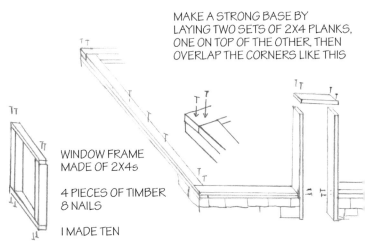

WINDOW FRAME MADE OF 2X4s

4 PIECES OF TIMBER 8 NAILS

I MADE TEN

The main structure is an idea borrowed from America. Get yourself a lot of wood called 2x4 (two by four) and some 3-inch nails. Your main beams are two 2x4 planks, nailed on top of each other. That enables you to make lots of joints. For instance, around the top of the brick wall, I laid my 2x4s in such a way that the underneath plank projected by two inches further out than the top one. This gives you a nice overlapping joint and the whole thing can be slammed together rapidly with the 3-inch nails.

I aimed for nails as often as possible as screws take more time.

Having laid a base around the wall, I made a set of eight identical frames, the same size, laying them on each other so each was the same size and I didn't need to bother with a tape measure.

When it came to the joints between the uprights and the roof beams, I cross-braced each upright and roof beam with a strip of mild steel with a hole through at each end and a bolt through the beam.

Again, doing it this way, you can do it by eye without needing to measure. This structure was strong enough to take my weight when I nailed on the roof tiles. I take no responsibility if yours isn't. Only making a suggestion.

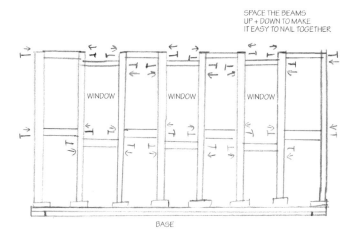

SPACE THE BEAMS
UP + DOWN TO MAKE
IT EASY TO NAIL TOGETHER

WINDOW   WINDOW   WINDOW

BASE

THE WALLS

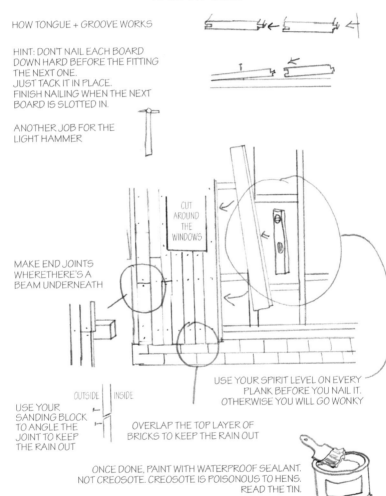

HOW TONGUE + GROOVE WORKS

HINT: DON'T NAIL EACH BOARD
DOWN HARD BEFORE THE FITTING
THE NEXT ONE.
JUST TACK IT IN PLACE.
FINISH NAILING WHEN THE NEXT
BOARD IS SLOTTED IN.

ANOTHER JOB FOR THE
LIGHT HAMMER

CUT
AROUND
THE
WINDOWS

MAKE END JOINTS
WHERE THERE'S A
BEAM UNDERNEATH

USE YOUR SPIRIT LEVEL ON EVERY
PLANK BEFORE YOU NAIL IT.
OTHERWISE YOU WILL GO WONKY

OUTSIDE | INSIDE

USE YOUR
SANDING BLOCK
TO ANGLE THE
JOINT TO KEEP
THE RAIN OUT

OVERLAP THE TOP LAYER OF
BRICKS TO KEEP THE RAIN OUT

ONCE DONE, PAINT WITH WATERPROOF SEALANT.
NOT CREOSOTE. CREOSOTE IS POISONOUS TO HENS.
READ THE TIN.

For the walls, I used tongue and groove, which is expensive but looks nice. See some hints on the left.

For the roof, I used split cedar shingles because they are supposed not to harbour insects. If they are overlapped properly, they are wonderfully dry, but it is a bit of an art to nail them on the roof whilst sitting there yourself especially at the edges, as there is nowhere to sit.

For the top ridge of the roofline, the cedar supplier did have some V-shaped pieces which I found fragile. If I had been less fussy, I would have nailed down another strip of damp-proof membrane to keep the rain out.

There's a certain architectural folly aspect to our henhouse as I found a Victorian galvanized iron chimney dumped in the garden of the kind that has a rotating windmill, fitted in old vans. In theory on hot days the windmill spins and sucks the hot air out from indoors.

I dismantled this and made a point bearing for the bottom of the windmill out of a snapped drill bit sitting in an oilbath and to my amazement, it still spins in the wind nine years later.

THE ROOF

START AT THE BOTTOM

9. IF YOU GET GAPS, CHEAT WITH A HIDDEN BIT OF DAMP-PROOF MEMBRANE

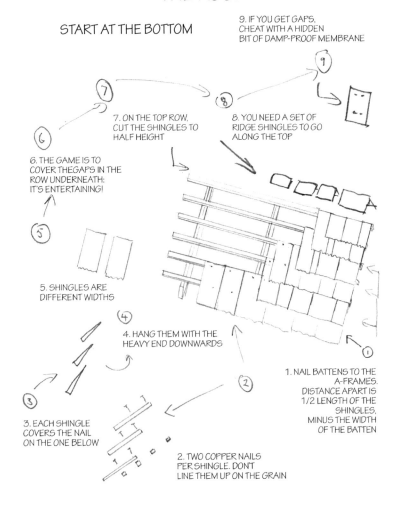

7. ON THE TOP ROW, CUT THE SHINGLES TO HALF HEIGHT

8. YOU NEED A SET OF RIDGE SHINGLES TO GO ALONG THE TOP

6. THE GAME IS TO COVER THE GAPS IN THE ROW UNDERNEATH: IT'S ENTERTAINING!

5. SHINGLES ARE DIFFERENT WIDTHS

4. HANG THEM WITH THE HEAVY END DOWNWARDS

1. NAIL BATTENS TO THE A-FRAMES. DISTANCE APART IS 1/2 LENGTH OF THE SHINGLES, MINUS THE WIDTH OF THE BATTEN

3. EACH SHINGLE COVERS THE NAIL ON THE ONE BELOW

2. TWO COPPER NAILS PER SHINGLE. DON'T LINE THEM UP ON THE GRAIN

The windows are glass rectangles set in home-made wooden frames. I fitted a single screw half way up the frame on each side and screwed it into the wooden post on either side, so that the window rotates. At the top of the window, I fitted a rotating wooden peg with a screw through it. This means that if you shut the window and turn the peg, it holds it firmly shut; if you open the window and turn the peg, it blocks the window to predators, but holds it open a couple of inches for ventilation only.

By using bits of wood in this way, you can save yourself the cost of a load of window furniture.

Getting in and out

The door belonged to a neighbour who threw it out. I made it into a stable door by cutting it in half, which proves very useful as Jane can look into her hens and still keep them locked in.

I also made a pop-hole for the hens, which is their real front door. This was made by cutting into the door with a jigsaw and ended up a charming gothic shape with a straight cut across the base but two curved sides to a point at the top.

Not because it's elegant but because I am lazy and you can make this door with three cuts not four! Plus one cut up the centre.

Having made the pop hole, we found that the dog could squeeze into it and go prospecting for eggs, so I had to make a ply inner door which is rather smaller. Where a dog can go, a fox can go.

Across the centre of the pop hole doors is a simple brass bolt. Brass is less likely to corrode than iron outdoor bolts.

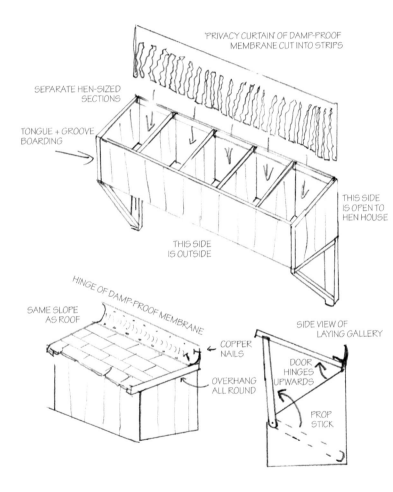

'PRIVACY CURTAIN' OF DAMP-PROOF MEMBRANE CUT INTO STRIPS

SEPARATE HEN-SIZED SECTIONS

TONGUE + GROOVE BOARDING

THIS SIDE IS OPEN TO HEN HOUSE

THIS SIDE IS OUTSIDE

HINGE OF DAMP-PROOF MEMBRANE

SAME SLOPE AS ROOF

COPPER NAILS

OVERHANG ALL ROUND

SIDE VIEW OF LAYING GALLERY

DOOR HINGES UPWARDS

PROP STICK

Laying gallery

With this design, there is always a problem of making the hinged section rainproof, so that water running down the side of the henhouse does not get in. The solution is a sheet of plastic called damp-proof membrane, very cheap.

Build it into the roof so it sticks out and pushes against the wall; it flexes as you lift the roof but means that the hinge does not let in water.

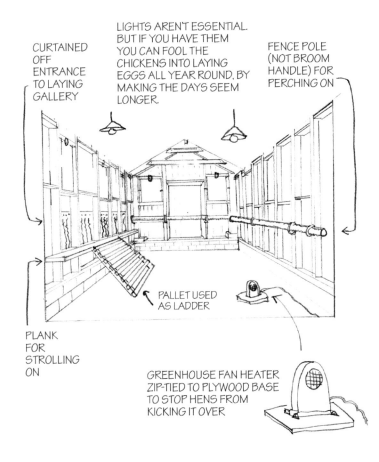

CURTAINED OFF ENTRANCE TO LAYING GALLERY

LIGHTS AREN'T ESSENTIAL. BUT IF YOU HAVE THEM YOU CAN FOOL THE CHICKENS INTO LAYING EGGS ALL YEAR ROUND, BY MAKING THE DAYS SEEM LONGER.

FENCE POLE (NOT BROOM HANDLE) FOR PERCHING ON

PALLET USED AS LADDER

PLANK FOR STROLLING ON

GREENHOUSE FAN HEATER ZIP-TIED TO PLYWOOD BASE TO STOP HENS FROM KICKING IT OVER

Inside comfort

We gave the hens a clear ledge at the entrance to the laying gallery so they can promenade before choosing which gallery to lay into – there are definite fashions which change!

The ladder leading to this ledge is made of an old pallet cut in half, and when it gets dirty, periodically we use it for firewood and replace it. Hens are surprisingly heavy. Pallets are better than a home-built ladder. You can find pallets easy and free to come by.

They also have a long pole made of a garden centre fencing pole. It must be wooden, about four inches in diameter, for their claws to grip. Plastic poles are unfair, as the hens slip as they sleep. Broom handles are too weak unless for only two three birds. The birds sleep in a row on the fence post and it is important.

Hens can be persuaded to carry on laying through the winter if you fit a henhouse light and leave it on till nine or ten at night, as long as you remember to turn it off.

For this purpose we fitted a waterproof switch on the outside of the henhouse.

We also fitted a couple of water-resistant sockets to the electricity supply so that we can use one for pressure washing the henhouse and the other for a greenhouse heater for the coldest nights. See Jane's earlier and important note – screw them or cable-tie onto a piece of ply so that the hens don't kick them over in the night, making it impossible for them to be too close to the wall or a pile of straw.

Having spent days lovingly making Jane's henhouse, it recently occurred to me that the classic French film shot of hens fluttering out of an old Citroen van, is quite a decent way to have an instant henhouse.

An old car with some hay or straw in the bottom, doesn't allow easy access to predators like rats or foxes, has ventilation, as long as you keep the windows open when it gets hot, and also windows for them to look out of and places to perch – the seats would get mucky quickly of course unless you took them out or covered them. Also you can move it around your land and rig up a covered run at the back if wanted.

Another idea is to adapt an old garden loo or shed as long as the base is concrete or stone or brick so that predators can't burrow up.

If you want to work with the very cheapest materials, old doors and pallets are easiest to get hold of. It is best to build a henhouse from planed wood because unplaned wood is like a lodging house for spider mites and it will make your hens' lives difficult. You will need to spend time with a belt sander, planing wood down.

Anyone with a bit of vision should be able to turn seven

or eight old doors into a respectable henhouse with two doors butted up against each other for the roof, and other doors used as you wish for the sides, front and back, filled in with damp proof membrane which solves most water problems.

This is the cheapest option, but make sure that you have a base. If you don't have a concrete base to hand, build the henhouse on legs lifting it three feet above the ground. This makes it hard for foxes to get in. It may disconcert hens to have foxes and rats running under the henhouse, but as long as the base is well built and lifted from the ground, the fox will be frustrated.

Put windows in it. A dark henhouse is miserable and hard to clean.

If you have real fox problems, try electric fences. They are becoming very popular apparently.

Afterword
by Will Tribble

with help from Andy and Charlie

I think this book of advice finishes a little suddenly. And so I will attempt to write an ending.

On 14 May 2012, I was sitting working on my mother's bed, when I noticed the sun was setting. It was time to put the chickens away. Boo had been sitting around and whining, holding up a single paw that the vet had repeatedly told us wasn't at all injured. So he was happy to run downstairs with me and help round up the clucking hens.

We put down some extra food for the flock in the shed and the rescue hens that were acclimatizing safely in the greenhouse. I picked up a few eggs that had been casually laid in the chicken shed during the day, I probably gave one to Boo. I fed the dog, then the cats, based on their own dietary requirements and strange neurotic preferences.

Then I went back upstairs. We had moved her bed so she could look out of the window and see the hens and dog milling and clicking about, as that made her truly happy. I watched my mother sleep for a little while. And then I realised that, while I was gone, she had stopped breathing.

And that was that.

It was not unexpected. We had known it was going to happen. The most bizarre thing about it was that life continued to unspool around us. There was no change to the texture of the air. Gravity continued to hold us to the ground. I replied to a work email by writing, 'I'm very sorry. I don't think I can make the meeting tomorrow or drafts for some time. My mother has died approximately 10 seconds ago'.

And then we went on to the next thing that needed doing, which was crying, and then the next, which was admin. Our house still was used for film and fashion shoots, even while Mum was ill: we just kept her room out of bounds. That evening, the undertakers took her away; the next day, a film crew were in the house for a shoot. We said nothing about it. Didn't want to spoil their rhythm. She would have smiled at that.

We even got upset at well-wishers bringing flowers, her ethos was she wanted the rectory to carry on, much as her energy carried us on. We continued doing whatever was immediately needed, one step in front of the other. And now we come to today, whatever day that is.

I am sorry to burden you with this at the end of a book of advice about henkeeping. Think of it as an extension of that work email, an apology for a delay. In some way Jane Furnival and hens are now intrinsically linked in my head. She had written her own obituary for the UK Press Gazette; but when a journalist called asking for more stories about my mother, all I could think of, playing over and over, was her catching that damn cockerel after he ran into the library. At least now I know why the chicken crossed the road.

Mum wanted to live her life by her own rules. Keeping chickens was only part of that. She also swam in crocodile infested waters in Botswana, rode a sea plane across the Caribbean, hung out with Ted Danson on the banks of Loch Ness, lived on a houseboat with a duck that said Thank You, and converted a Gothic confessional into a functioning toilet. She became a local celebrity not just through her writing and broadcasting, but through the sheer power of her presence, a ball of energy bouncing around with a large fluffy dog. All of us – me, Charlie, Henry, Dad – were happy to be part of this adventure. We cared for the hens because we cared for her. And after she and Boo died, rather than a gravestone we got her a comfortable bench in Nonsuch park, not far from where we once lived in Cheam. You're welcome to have a look for it. It's a lovely walk.

I feel as though many of my additions to this book have been more downbeat than upbeat: don't do this, you can no longer do that, that's definitely illegal... So let me take a negative beginning and try to turn it into a positive end.

Chickens are short lived creatures, with many predators, whose wings do not enable them to fly. You will need to take care of them but also you will need to protect them. You have to keep an eye out for them, and listen to make sure they're clucking, not squawking. You must accept that they cannot be trained not to make a mess of things. You have to know that they will sometimes fight you even as you try to help them. You will surround yourself with funny little animals that jump on your knees when you're reading, scamper behind you as you walk, peep at you through your windows. You will have a reason to wake up early and an excuse to be outside every sunset. You will have friends come to see and children that you must teach to be gentle. You will always have gifts to give, warm and fresh, to those that can appreciate them. You will have life, fluttering around you, with its beauty and ugliness, comedy and tragedy. And as you reach in your fridge the next time you make a cake, or an omelette, or just some nice breakfast, you can think, 'I helped this happen'.

And that is reason enough for keeping hens.

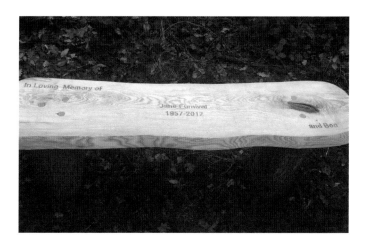

Knowledge and Resources

The following links have all been checked and updated as of summer 2020.

The British Hen Welfare Trust

Hope Chapel, Rose Ash, South Molton, Devon EX36 4RF
01884 860084
info@bhwt.co.uk
www.bhwt.org.uk

The main resource for information about keeping hens, adopting a rescue hen, and all other enquiries. Their hen helpline is the same as above, and they have a robust health advice section. They also accept donations.

Don't want hens at home? You can sponsor a hen through the BHWT.

Legbars of Broadway

www.legbarsofbroadway.co.uk

Philip Lee-Woolf invented the Cotswold Legbar blue egg-laying hen and is the fount of all knowledge on so many things, plus you can become a member of the Legbar hen club and get discounts and also install a webcam looking at their hens on your computer.

Southmead Poultry

The Glade, Fetcham, Leatherhead KT22 9TF
01372 458320 | 07963 013597
www.southmead-poultry.co.uk

Recommended hybrid chicken seller.

Tracy of Southmead Poultry adds that 'we also sell chicken coops, poultry accessories, in fact everything you need to keep chickens, and we offer a hen boarding service too (for when you go on your holidays and need someone to look after your hens).'

Anthony Allen, Cotswold Chickens

Edgehill Stud, Stratford Road, Kineton, Warwickshire, CV35 0DX
01608 683912 | 07779 263296
www.cotswoldchickens.com

All hen supplies including fancy breed chickens.

The Poultry Club have a directory of specialist breeders.

www.poultryclub.org

Feeding

Small Holder Range from **Allen and Page**

www.allenandpage.com/the-smallholder-range/

Gastro Grit from **The Little Feed Company**

shop.bhwt.org.uk/collections/chicken-treats/products/gastro-grit

Metal kitchen bins for chickenfeed can be found in multiple sizes at Habitat or Garden Trading.

www.habitat.co.uk
www.gardentrading.co.uk

Manuka honey

www.manukahoney.co.uk

Bedding

Dust-extracted straw bedding by **Helmes**

http://www.regencyfeeds.co.uk/Helmes-Straw

Dengie Fresh Bed is highly recommended

shop.bhwt.org.uk/collections/bedding/products/dengie-fresh-bed-50ltr

Henhouses

B&Q

www.diy.com

Forsham Cottage Arks

www.omlet.co.uk

Top quality plastic housing, plus they sell an automatic pop hole with a timer but it's expensive - upwards of £125 - and something new called 'the pullit' in the line of automatic door openers.

Flyte so Fancy

www.flytesofancy.co.uk

Top quality wooden housing.

Green Frog Designs

www.greenfrogdesigns.co.uk

Top quality recycled plastic housing.

Books

Keeping a Few Hens in Your Garden
by Francine Raymond

www.kitchen-garden-hens.co.uk

The Secret Life of Cows
by Rosamund Young

A fantastic book by a working farmer who discusses the personalities of animals and opens your eyes to the way they think and why they do things in a very easy to read style. Cheapest used copy from Amazon, £4.25.

For a wide range of smallholding books:
www.firstpasture.co.uk

Information

Henkeepers Association

www.henkeepersassociation.co.uk

Online information network for henkeepers.

The Poutry Pages

www.chickens.allotment-garden.org

Chicken Talk

www.chickentalk.wordpress.com

A chatty blog from a committed hen lover.